Book 12
Probing The Planets

Contents

Eric Einstein finds a new use for a USO (unidentified space object)

Introduction

Probing The Planets

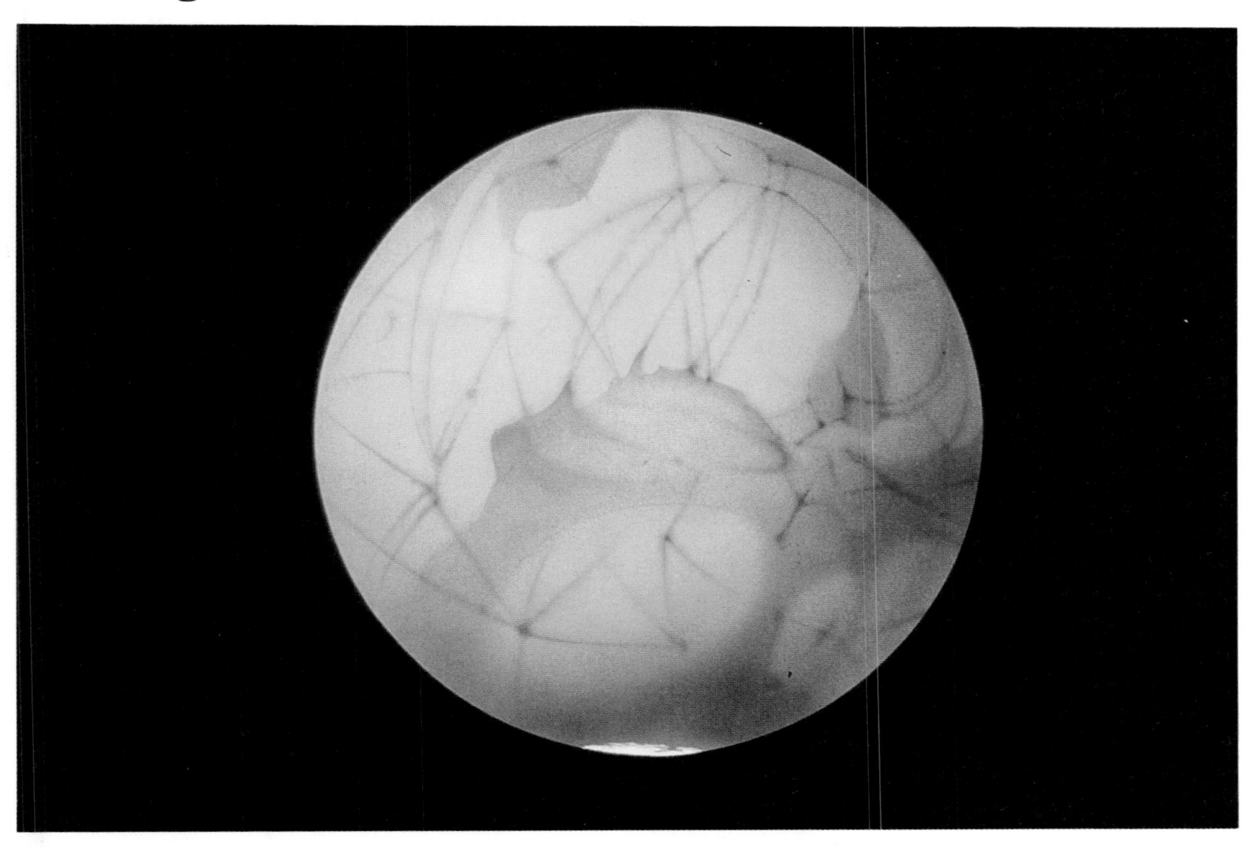

Do You Know...

- how we know what it is like on other planets?
- what the four planets made of rock are like?
- what the four planets made of gases are like?
- what Pluto, made of ice and rock, is like?

Chapter 1 Looking, Looking, Looking

■ How do people get information about the planets?

If you want to learn about rocks, you look both on and under the ground. If you want to study clouds, you must look at the sky. If you want to see a planet, you also must look at the sky. Unless, of course, you want to see planet Earth. Then you can simply look down at the ground.

By looking at the sky, you can sometimes see five planets. At night, you might see Mars, Jupiter, and Saturn. In the early morning or evening, you might see Mercury and Venus. In fact, Venus is so bright you can sometimes see it during the day.

From top: Mercury, Mars, Saturn, Venus, Jupiter

What can you observe in the sky? (Apart from your pointing finger!)

Telescope and observatory

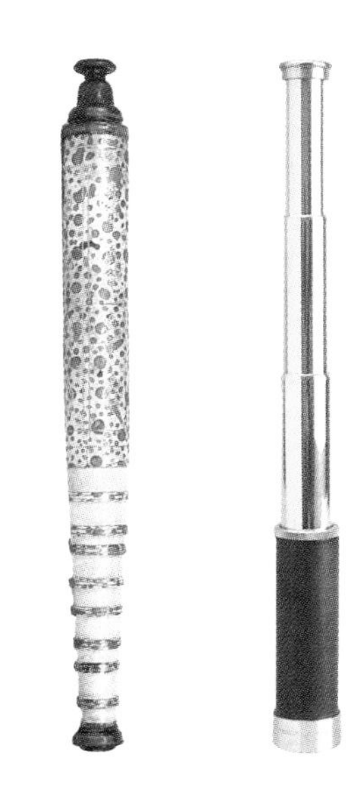

Old and new simple telescopes

Observing the Planets from Earth

A **planet** is a very large object that moves around the sun. A planet's position in the sky changes as it moves around the sun. You will not find a planet in the same place night after night. To see a planet, you must know where to look. Because planets seem to move among the stars, they are called the wanderers.

If you looked with just your eyes, a planet would look like a star. With just your eyes, you could not see very much at all. But you could learn more about planets by using a telescope. A **telescope** is a special instrument that makes faraway objects look closer and larger. With a small telescope, you could see that Saturn has rings.

Yeah! I'm the wanderer.

Eric Einstein's fantasy-wandering with the planets.

Observing Planets From Space

Some planets are too far away to see with even a telescope. To learn more about these planets, scientists send probes into space. A **probe** is a spacecraft that explores space and other planets.

Space probes provide several kinds of information. They can fly by a planet and take pictures. They can land on a planet and sample the soil. They can also study a planet's weather and atmosphere.

Two probes, Voyager 1 and Voyager 2, left Earth by September 1977. From them, people have learned much about four planets very far from Earth.

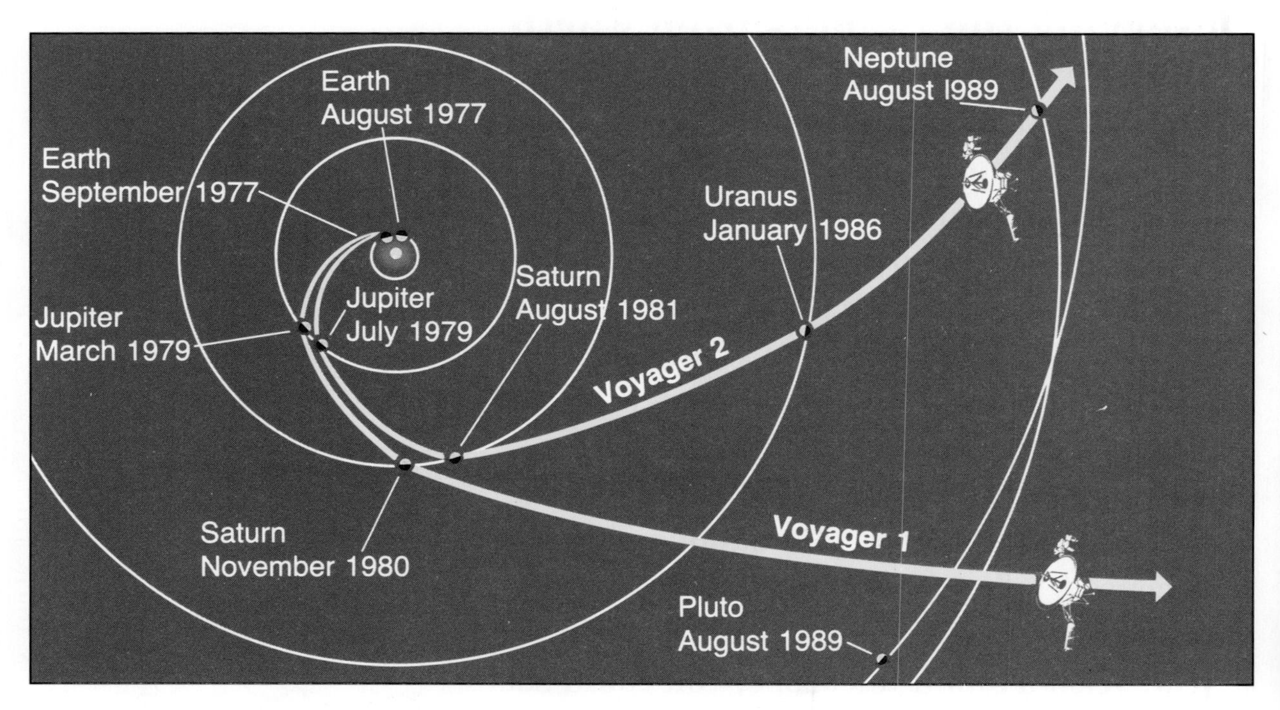

Voyager space probes sailing into the unknown

Into the Black Abyss!

Voyager 2 passed Neptune on 25 August 1989. It is now heading to the heliopause — the place between the solar system and the stars — where no spacecraft has been before. It has enough power and fuel to work until the year 2015 but it will still drift on after that. In about 296,000 years it will pass Sirius, the brightest star in our sky.

Chapter 2 The Rocky Ones

■ What are the four planets closest to the sun like?

Mercury, Venus, Earth, and Mars are the four planets closest to the sun. They are called the inner planets. All four of these planets are made of rock. For that reason, they are also known as the **rocky planets**.

Mercury The planet Mercury is closest to the sun. It is less than half the size of Earth. Unlike Earth, Mercury has no atmosphere.

Mercury rotates or spins only once every 59 days. On Mercury, the temperature reaches 427°C. That is hot enough to melt lead.

Space geologist in search of hot rocks.

Mercury — top hot spot!

The surface of Mercury is covered with craters and giant cliffs. It also has large flows of volcanic lava, just like the surface of Earth's moon.

Venus The planet Venus is the brightest in the sky. It is almost the same size as Earth. Venus, however, rotates only once every 243 days. Temperatures on Venus reach 470°C.

Venus is surrounded by a thick, cloudlike atmosphere. For this reason, you cannot see the surface with a telescope. The clouds also reflect sunlight. This is why Venus is so bright. Space probes have found mountains, volcanoes, and large craters on Venus.

Venus, Goddess of Love and Beauty, steals the solar show.

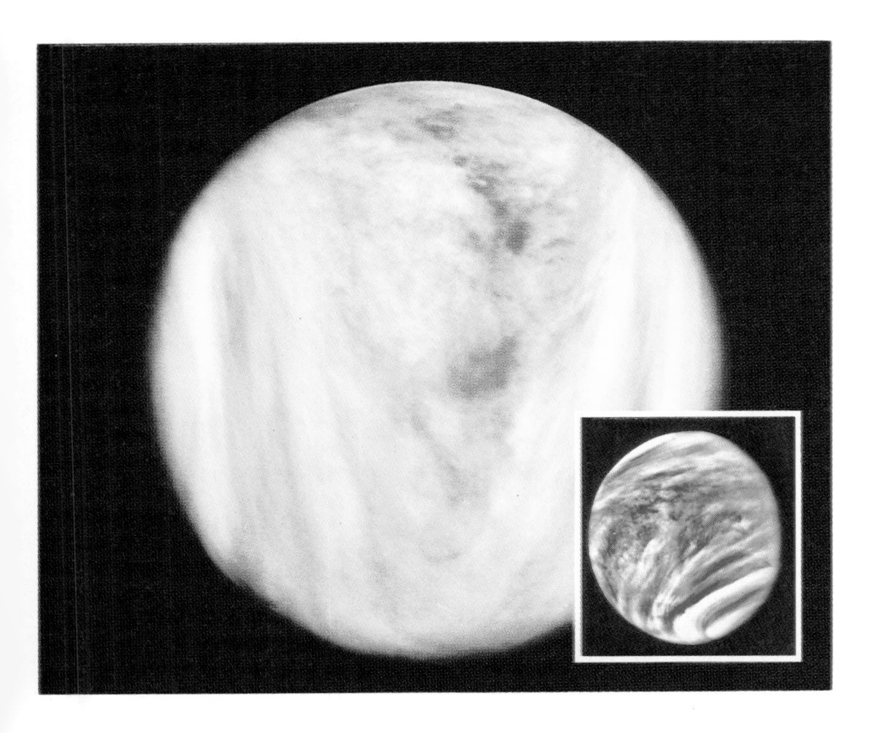

Two shots of Venus

Earth seen from surface of moon

Earth The planet Earth is called the water planet. It is the only planet known to have water on its surface. Almost three fourths of Earth's surface is covered by water. Earth is also the only planet in our solar system known to have living things on it.

Unlike other planets, Earth has an atmosphere that can support life. Earth's atmosphere also produces weather. At the Earth's poles, there are icecaps. The temperature can be as low as –34°C at the North Pole. At the equator, the temperature may get to more than 49°C during the day.

Earth's crust is rocky. The land masses are covered by a mixture of rock and soil. There are high mountains and flat, low areas. Water sometimes cuts deep canyons and gorges into Earth's rocky crust.

Mars The planet Mars is about half the size of Earth. Mars rotates once every 24 hours and 36 minutes. Therefore, its day is just slightly longer than a day on Earth.

Mars has an atmosphere that is much thinner than Earth's. As a result, atmospheric pressure on the surface of Mars is 100 times less than the atmospheric pressure on the surface of Earth.

Mars is a cold planet. At its equator, daytime temperatures are about –17°C. During the night, temperatures drop to about –130°C.

Eric Einstein loses his love for rocks.

Mars and its surface — rocks, rocks and more rocks

Red, rocky desert dweller.

The surface of Mars is covered with craters and with desert-like regions. It also has features that look like dry river beds. There are four giant volcanoes on Mars. There is also a gorge so long that it could stretch across Australia.

Large patches of the surface of Mars look red, because the soil contains rusted minerals. For this reason, Mars is sometimes called the red planet.

More Data About the Rocky Planets

Planet	Mercury	Venus	Earth	Mars
Distance from sun	58,000,000 km	108,000,000 km	150,000,000 km	228,000,000 km
Time of revolution (Earth time)	88 days	225 days	1 year	2 years
Diameter	5,000 km	12,000 km	13,000 km	7,000 km
Probes that provided data	Mariner 10 (1974, 1975)	Mariner 2 (1962) Venera 4 (1967) Mariner 5 (1967) Mariner 10 (1974) Pioneer Venus 1 and 2 (1978) Venera 13 and 14 (1982)	Explorer XVII (1963) Landsat 1 (1972) Nimbus-7 (1978) GOES-D (1980) Many others	Mariner 4 (1965) Mariner 6 and 7 (1969) Mariner 9 (1971) Mars 4 and 5 (1973) Viking 1 and 2 (1975)

What's in a Planet's Name?

Mercury — messenger of the Roman gods, who wore a winged helmet and winged sandals. He was also god of science, commerce, patron of travellers, rogues, vagabonds and thieves!

Venus — the Roman goddess of love and beauty, and mother of Cupid. Her symbol is the mirror. On April 1, there were festivals to honour her...

Earth — there is no god or goddess called Earth, but the ancient Greeks worshipped a goddess of the Earth. Her name was Gaea. She gave birth to sky, mountains and sea.

Mars — the Roman god of war, and usually seen as an armed warrior who also protected the ancient city of Rome. He started out life as a god of agriculture!

Jupiter — the greatest of all the Roman gods. He was lord of the heavens and the bringer of light. White was his sacred colour. His statue at Olympia was one of the Seven Wonders of the World.

Saturn — the Roman god of agriculture. His symbol is the scythe. He was the father of all the gods. A festival in his honour was held in December — it has been taken over by Christmas. His name lives on in the word 'Saturday'.

Uranus — a Greek god and husband of Gaea. He hated his children and hid them in the earth. His son Kronos rose up and defeated him.

Neptune — the Roman god of the sea who always carried a trident and rode around his watery world on a dolphin. He was a stately old man with a long white beard.

Pluto — the Roman god of the underworld, place of the dead. The Greeks called him Hades. He stole the beautiful Persephone while she slept and made her queen of the underworld. Once a year he allows her to return to the earth. When she does, she brings the spring.

Chapter 3 The Frozen Ones

■ What are the five planets farthest from the sun like?

Another kind of space probe.

Jupiter, Saturn, Uranus, Neptune, and Pluto are the five planets farthest from the sun. They are called the outer planets. All five are **frozen planets**. All but Pluto are thought to be made up mostly of gases. Pluto is probably made of ice and rock.

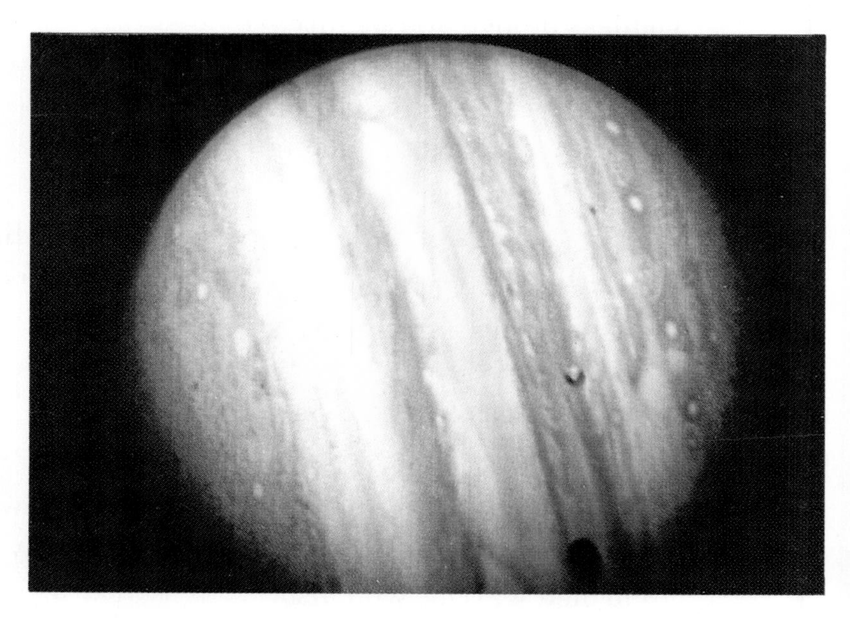

Jupiter, the largest planet, battles with its bulge!

Jupiter Jupiter is the largest planet. It is 11 times bigger than Earth. Its atmosphere is cloudy and very cold. At the top of the clouds, its temperature is about –130°C.

Jupiter spins rapidly. Even though it is the largest planet, it takes only ten hours to rotate. This rapid spinning causes Jupiter to bulge at its equator. The fact that Jupiter is not solid also makes it bulge.

For a long time, scientists were puzzled about a red spot on Jupiter. Probes flew close to Jupiter in 1974 and 1975. They sent back information about the red spot. Scientists now think this red spot may be a huge storm. It may have been raging for hundreds of years.

Jupiter's red spot — a blot on the face of a giant?

Saturn The planet Saturn is the second largest planet. It is about nine times bigger than Earth.

Saturn is like Jupiter in many ways. Both planets are made mostly of gases. Both planets have atmospheres that are very cold. The temperature at Saturn's cloud tops is about –185°C. Both planets spin rapidly. Saturn rotates once every 17 hours.

Eric Einstein falls in love with a beautiful planet.

Saturn and three moons

Saturn is well known for its rings. Probes have sent back interesting pictures of these rings. Scientists learned that the rings are made of chunks of rock and ice. Some chunks are as small as a basketball. Some chunks are as large as a small car.

Uranus Scientists think Uranus may have an atmosphere, crust, mantle, and core. The planet rotates in 13 hours. It is about four times bigger than Earth. Recently, scientists learned that Uranus has rings.

The planet Uranus is tipped over on its side. This means the north pole of Uranus points toward the sun. When the sun shines on its north pole, there are 42 years of sunlight there. The temperature at Uranus' cloud tops is about –216°C.

Two shots of Uranus

Neptune as pictured from one of its moons

Alberta Skywalker discovers Neptune on the rise.

Neptune The planet Neptune is probably a twin of Uranus. But Neptune is a little smaller than Uranus. And the matter making up Neptune is more dense (thicker) than the matter making up Uranus.

Neptune rotates once every 18 hours and 12 minutes. Its atmosphere contains a thick layer of clouds. The temperature at Neptune's cloud tops is about –200°C. Like Earth, Neptune has seasons. However, temperatures at any season of the year are much colder on Neptune than on Earth.

Pluto Pluto is a small planet. It is less than half the size of Earth. Pluto rotates once every six days and eight hours. The matter making up Pluto is only slightly more dense than water.

Surface of Pluto

Maybe it's time to get these guys out of the deepfreeze!

Very little is known about Pluto because it is so small and so far away. It must be very cold and very dark. The temperature at the cloud tops is about –150°C. From Earth, Pluto can be seen only with a very powerful telescope. Even then, it looks like a point of light.

More Data About the Frozen Planets

Planet	Distance from sun	Time of revolution (Earth time)	Diameter	Probes that provided data
Jupiter	779,000,000 km	12 years	143,000 km	Pioneer 10 (1973) Pioneer 11 (1974) Voyager 1 and 2 (1979)
Saturn	1,428,000,000 km	29 years	121,000 km	Pioneer 11 (1979) Voyager 1 and 2 (1980 and 1981)
Uranus	2,871,000,000 km	84 years	51,000 km	Voyager 2 (1986)
Neptune	4,499,000,000 km	165 years	45,000 km	Voyager 2 (1989)
Pluto	5,915,000,000 km	248 years	3,000(?) km	None

Beyond the Naked Eye — Mystery Planet 'Y'?

Did you know that Uranus was not discovered until 1781? Sir William Herschel spent many long, lonely nights staring through his telescope at the inky blackness of the sky. His work was rewarded. He found Uranus.

Eric Einstein goes tracking for planets.

It wasn't until 1930 that the tiny, distant and frozen world of Pluto was discovered. Clyde Tombaugh had taken up the search made by early astronomers. They were certain there was a ninth planet and had named it Planet X. Because it's just so far away, its only moon, Charon, wasn't discovered until 1978.

Eric Einstein in disguise as a maths genius, tracks down Planet Y.

Eric Einstein observes the stars on the ceiling.

The giant planet Neptune was not discovered until 1846. A German astronomer called Johann Galle made this startling find. Thanks to the work of others and their wonderful mathematics, he was the one who finally tracked Neptune down.

The astronomer's nightmare

Now, some astronomers feel there may be a tenth planet in our solar system. Perhaps there is, perhaps there isn't. Shall we call it Planet Y? Nah! You can think of something better can't you?

Chapter 4 Technology Today

Probe photo of cyclone off Queensland

Landsat-C probe

Using Pictures from Probes

Scientists can learn many important things from pictures taken by probes. For example, weather probes have cameras that sense heat. Normal cameras cannot take pictures of clouds at night. These special cameras can. Therefore, a probe can observe a hurricane on Earth at night.

These special cameras can also be used in other ways. In their pictures, healthy plants on Earth look bright red. Therefore, scientists can quickly spot large areas of diseased plants. Then the scientists can act to prevent the disease from spreading.

Special computers can change the colours in a picture. The new colours can show things scientists might otherwise miss. For example, colour can be used to find the source of pollution in a river on Earth.

Think about It

Suppose you are planning a probe that will carry special cameras. The cameras can sense sources of heat on a planet's surface and in its atmosphere. Where would you send the probe? What kind of pictures would you have the probe take? How would you use the information gathered?

Don’t Junk My Space!

Did you know that...

- When a rocket is launched into space it has to travel at 11 km per second to break through the Earth’s gravity.
- To achieve this power, a rocket needs three firing stages. As each stage burns out, the equipment falls away from the rocket to become space debris.
- Today, scientists believe there are over 7000 bits of debris floating about in space. Some are the size of baseballs, some are much larger. All of it is hazardous. Now we have to design space robots to go and collect the junk! Well done, Earthlings!

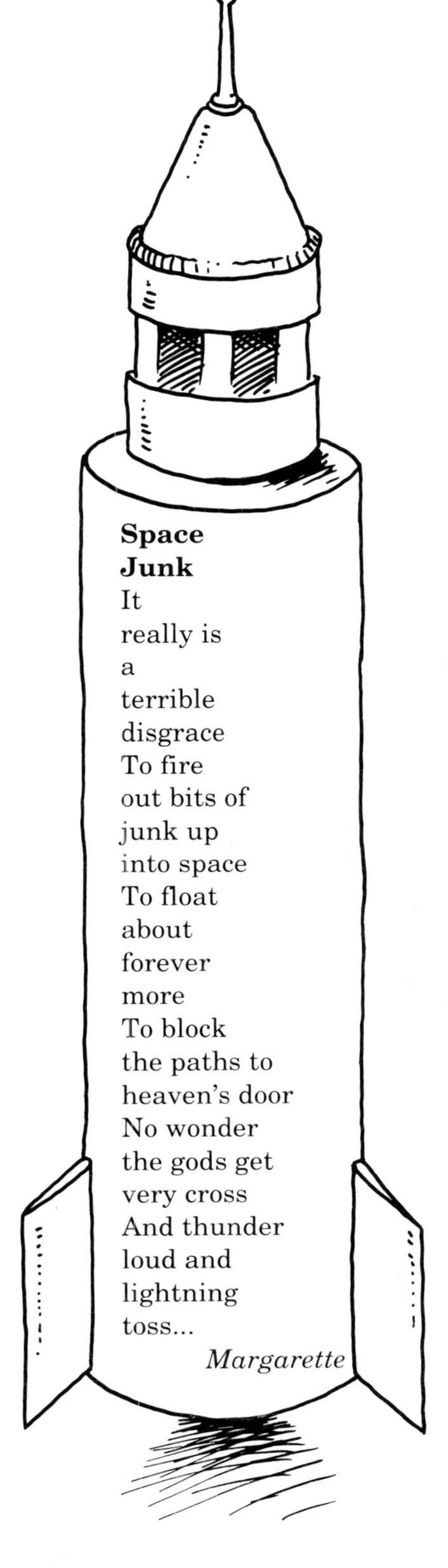

Check Up

Alberta Einstein's Brain Tester

Summary

- Telescopes and probes help us know what other planets are like.
- The rocky planets, or inner planets, are closest to the sun and are made of rock.
- The frozen planets, or outer planets, are farthest from the sun and are made up mostly of gases.
- Pluto, a small planet made of ice and rock, is probably very cold and dark.

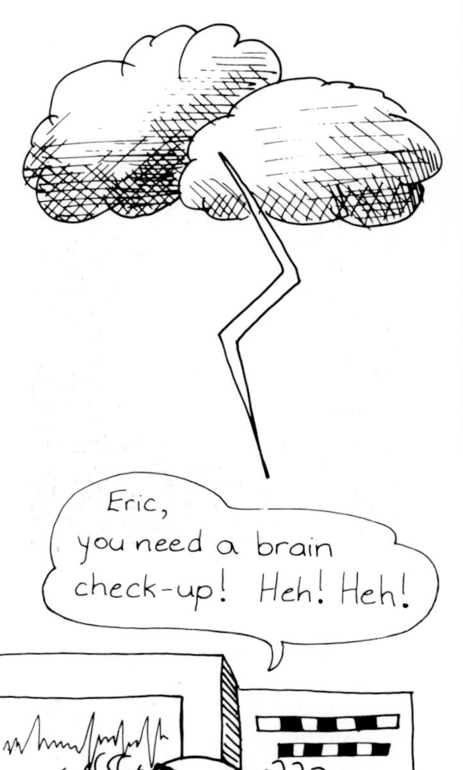

Scientist Ideas

1. Make a list from **a** to **d**. For each letter, write the name that does not belong.
   - **a.** Rocky planets: Venus, Earth, Jupiter, Mars
   - **b.** Made mostly of gases: Mercury, Uranus, Jupiter, Saturn
   - **c.** Frozen planets: Saturn, Pluto, Venus, Neptune
   - **d.** Planets: Jupiter, Neptune, Saturn, Sun

2. Make a list from **a** to **i**. Write the correct term for each letter in the diagram.

3. Look at the picture of Saturn. How could we get such a picture?

4. Data Bank
Use the table below to answer the following questions.
1. If you can jump 1 metre on Earth, how high could you jump on Mars?
2. What is the difference in height between jumps on Mercury and Jupiter?

Comparative Jumping Heights on Planets

Planet	Gravity	Height jumped (in centimetres)
Jupiter	2.34	43
Neptune	1.18	84
Uranus	1.17	85
Saturn	1.15	87
Earth	1.00	100
Venus	0.88	114
Mars	0.38	263
Mercury	0.37	270

Don't Gloss over the Glossary!

dense — thick
frozen planets — the planets farthest from the sun i.e., Jupiter, Saturn, Uranus, Neptune and Pluto
observe — watch closely
planet — a very large object that moves around the sun
position — a place
probe — search/explore
rocky planets — the planets closest to the sun and made of rock e.g., Mercury, Venus, Earth, Mars
rotate — to spin or turn on the spot
telescope — a special instrument that makes faraway objects look closer and larger
wanderers — a nickname for the planets because they seem to wander/move among the stars